D1515136

Around and About

Into space

Anita Ganeri
and Jakki Wood

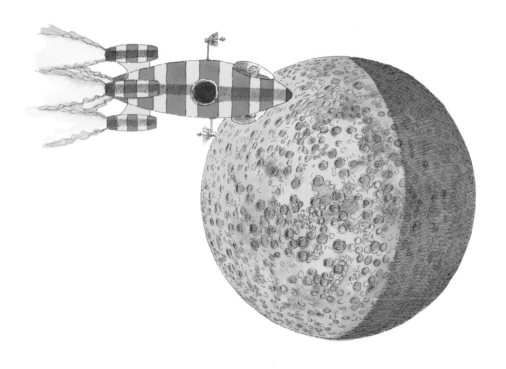

Barron's

First edition for the United States, Canada, and the Philippines published 1993 by Barron's Educational Series, Inc.

© Copyright by Aladdin Books Ltd 1993

Designed and produced by
Aladdin Books Ltd
28 Percy Street
London W1P 9FF

All inquiries should be addressed to:
Barron's Educational Series, Inc.
250 Wireless Boulevard
Hauppauge, NY 11788

International Standard Book No. 0-8120-1761-7

Library of Congress
Catalog Card No. 93-4083

Library of Congress Cataloging-in-Publication Data

Ganeri, Anita, 1961-
  Into space / Anita Ganeri and Jakki Wood.
    p. cm. – (Around and about)
  Includes index.
    Summary: Harry and Ralph visit the Moon, the planets, and a comet in their spaceship and discuss the stars, the Milky Way, and the possibility of life on other planets.
  ISBN 0-8120-1761-7
  1. Space flight–Juvenile literature. 2. Interplanetary voyages–Juvenile literature. 3 Outer space–Exploration–Juvenile literature. [1. Space flight. 2. Interplanetary voyages. 3. Outer space–Exploration.] I. Wood, Jakki. II. Title. III. Series.
TL793.P325 1993
520–dc20   93-4083   CIP   AC
Printed in Belguim
3456 4208 987654321

Design David West Children's Book Design
Illustrator Jakki Wood
Text Anita Ganeri
Consultants Eva Bass Cert. Ed., teacher of geography to 5-8 year-olds
Sue Becklake, author of many science books for children

Contents

This is planet Earth

One night, Harry and Ralph are looking at the stars through their telescope. They want to learn more about what is out in space.

Mercury Mars Uranus
Venus Jupiter Neptune
Earth Saturn Pluto

Our planet Earth is part of what is called the Solar System. There are nine planets in the Solar System. They are Mercury, Venus, Earth, Mars, Jupiter, Saturn, Uranus, Neptune, and Pluto. Can you find them all on Harry's chart?

The planets travel in giant rings around the Sun. Some of the planets have moons, like the Moon you can see from Earth.

Blast off!

Harry and Ralph usually travel in their hot air balloon but for this trip they need a spaceship. They strap themselves into their seats. Then, it's 5...4...3...2...1...0 and ... blast off!

It takes their spaceship about ten minutes to reach space. It has to travel very, very fast. Otherwise, the Earth's gravity would pull it back down.

Gravity is an invisible force that pulls things down towards the Earth, the Moon, or the other planets. Throw a ball up into the air. What happens to it? Gravity pulls the ball down to Earth.

It's very dark and cold in space. It's also very quiet because there is no air to carry sounds from place to place.

Harry and Ralph can float around inside their spaceship. This is because the Earth's gravity is much weaker in space.

Moon mission

The first stop on their journey is the Moon. Harry and Ralph put on their space suits. These will keep them warm and give them air to breathe, since there is no air on the Moon.

Harry and Ralph bounce along the surface of the Moon. They can see Earth in the sky. They feel much lighter on the Moon because its gravity is not as strong as Earth's gravity.

Look at these footprints! Someone's beaten us to it!

Harry scrambles down into a huge crater. There are hundreds of craters on the Moon. They were made millions of years ago by space rocks crashing into the surface.

They are Neil Armstrong's footprints from 1969. There's no wind to blow them away so they'll last forever.

From Earth, you can see dark patches on the Moon. These are huge, flat plains.

new moon   quarter moon   half moon   gibbous moon   full moon

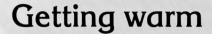

Getting warm

Harry and Ralph fly off towards the Sun. They cannot get too close. The Sun is so hot it would melt their spaceship.

The Sun is a star, right in the middle of the Solar System. There are millions and millions of other stars in space. But the Sun is the most important star for earthlings. It is huge compared with the Earth. The Earth gets all its heat and light from the Sun. Without it, the Earth would be too cold and dark for anything to live on.

The Sun's light takes eight minutes to reach Earth.

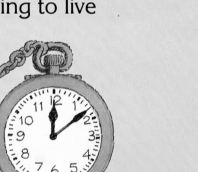

Never, ever look straight at the Sun. It will damage your eyes very badly.

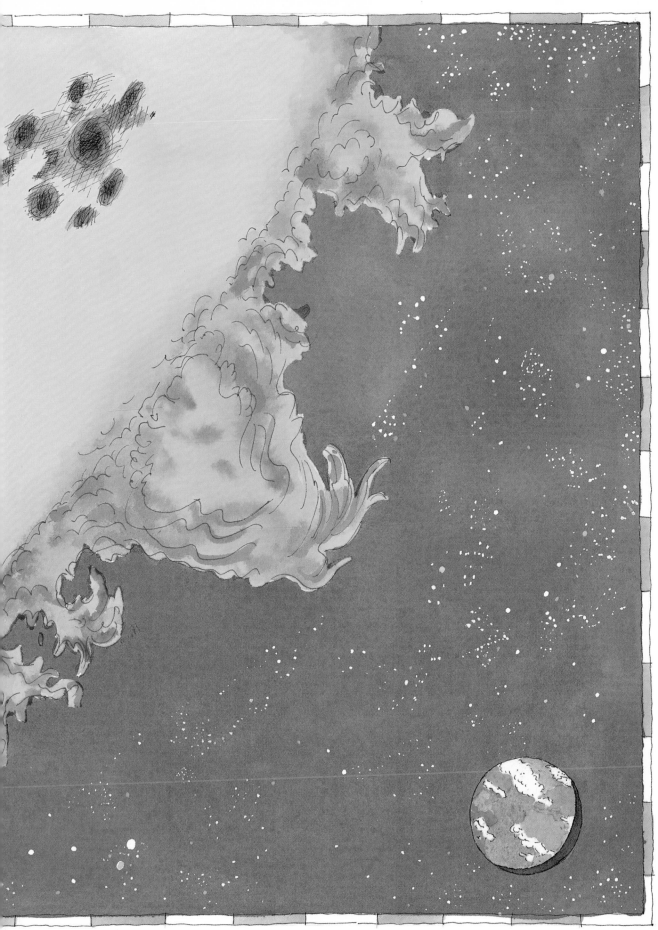

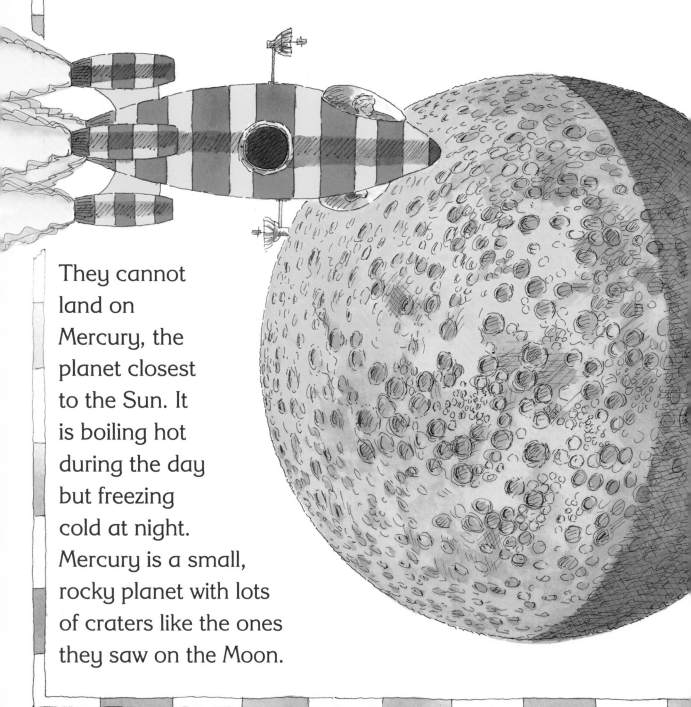

Mercury and Venus

Harry and Ralph turn their spaceship away from the Sun. They want to explore Mercury and Venus, the two planets closer to the Sun than Earth is.

They cannot land on Mercury, the planet closest to the Sun. It is boiling hot during the day but freezing cold at night. Mercury is a small, rocky planet with lots of craters like the ones they saw on the Moon.

The air is so thick that it will squash us if we try to land.

Let's keep our distance then!

Venus is a very different planet. Thick clouds of poisonous gas hide its dry, rocky surface. The clouds trap heat, making Venus much too hot to land on. Venus is the brightest of all the planets.

Life on Mars?

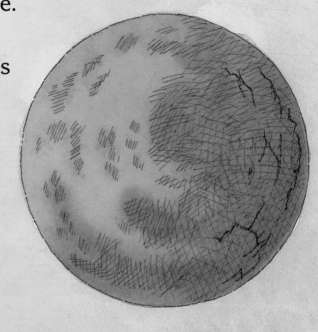

Mars is called the red planet because of its red, rocky surface and pink sky. Harry and Ralph have heard many stories about Martians but they cannot see any signs of life.

The North and South poles of Mars are icy. But most of Mars is dry and dusty. Rivers once flowed across its surface but they dried up millions of years ago.

They also explore a huge valley, called the Mariner Canyon. It is bigger than any Earth valleys. It would take more than a month to walk from one end to the other. Harry and Ralph don't try to!

They see a huge volcano. It is almost three times as high as Mount Everest on Earth.

I hope it doesn't erupt. Imagine how loud the roar would be!

Don't worry. The last time it erupted was over ten million years ago!

Jupiter and Saturn

On the way from Mars to Jupiter, Harry and Ralph fly past thousands of lumps of rock and metal. These are asteroids. Some are smaller than soccer balls. Others are as big as whole countries.

Harry and Ralph cannot land on Jupiter or Saturn. Both of these planets are made of gas and liquid. They don't have solid ground to walk on.

Look at that giant pizza!

That's Io, one of Jupiter's moons. There are fiery volcanoes all over its surface.

Jupiter is the biggest planet in the Solar System. It is 11 times as wide as the Earth. The gigantic red patch in its clouds is a huge storm.

Saturn is a beautiful sight with its shining rings. They are made of millions of pieces of ice all spinning around the planet.

Earth has only one Moon. But Jupiter has 16 moons and Saturn has 18 moons.

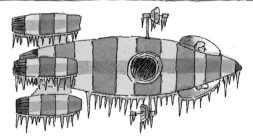

The outer planets

Harry and Ralph fly on to visit the last three planets in the Solar System. They are so far away from the Sun that they are very cold all the time.

Neptune

Uranus

Uranus and Neptune are gassy planets. The gases make them look bluish-green. They both have rings around them, and moons, too. Uranus has 15 moons and Neptune has eight.

Pluto is a rocky planet but it is far too cold to land on. In fact, Pluto is the coldest of all the planets, much colder than the coldest winter on Earth. Brrr!!! It is also the smallest planet.

Pluto has one moon, called Charon. It is bleak and icy.

Star struck

Harry and Ralph have already seen one star – the Sun. But there are millions of other stars in space. No one has been able to count exactly how many.

Stars are huge balls of glowing-hot gases, that make heat and light. This is why they shine. Planets don't make their own light, they just shine by reflecting sunlight.

The Sun is the closest star to Earth. It is 93 million miles (150 million kilometers) away. The next nearest star is Proxima Centauri. It is much further away than the Sun. The light we see coming from this star began its journey through space more than four years ago!

Look for patterns in the stars at night.
Can you find the Dog Star and Orion the hunter?

The Milky Way

A galaxy is an enormous cloud of stars. Our Sun and Solar System are just a tiny part of a galaxy called the Milky Way. There are about 100 billion stars in the Milky Way. And there are about 100 billion galaxies in space!

So, the Solar System is just a tiny speck in the Milky Way...

It's difficult to see the shape of the Milky Way from Earth. But Harry and Ralph have flown above the Milky Way. From here they can see that it is a spiral of stars.

...and the Milky Way is just one of billions of galaxies in space.

X Solar System

Comet ahead

As Harry and Ralph fly back towards Earth, they see a comet streaking past the Sun. A comet is a ball of ice, dust, and gas. When a comet flies near the Sun, the dust and gas stream out from it in a long, shining tail.

One comet had a tail long enough to wrap around the Earth more than 8,000 times.

Mine's just long enough to wrap around me!

Shooting stars, or meteors, look a bit like small comets. They are glowing streaks made when small pieces of space dust burn up as they fall through the air. You can sometimes see whole showers of shooting stars.

A safe landing

It is time for Harry and Ralph to come back down to Earth. They land safely. It has been a very long journey!

They think about all the stars and planets they have seen.
They wonder about life on the other planets.
They didn't meet any aliens on their journey. But space is such a big place, there must be someone else out there, somewhere!

Harry and Ralph draw some star constellations in their space scrapbook and name them. Next they want to find out how the planets got their names.

Index

This index will help you to find some of the important words in the book.